ANDREW ALEXANDER

6th Ave. & 19th St., New York City

Shoes of all Kinds and Styles

DEPARTMENT C

Men's and Boys' Shoes

Andrew Alexander

6th Avenue and 19th Street

New York City

Shoes of all Kinds and Styles

Department C

Men's and Boys' Shoes

NOTICE.

❧ ❧

T HIS CATALOGUE represents such goods as are kept constantly in stock.

Please read carefully the following in regard to ordering goods. Our **Terms** are **Cash.** Remit by Express, Money Order Draft, P. O. Order or Registered Letter, otherwise we forward by express, C. O. D., which last always entails additional expense.

When remitting, state amount enclosed. **No goods by mail or express sent on approbation or subject to inspection.**

We execute orders on the day received, or at the latest the following day, otherwise an acknowledgment will be mailed explaining the cause of delay ; but, should you not receive a reply within a reasonable time, please duplicate order, and give particulars of the enclosures in original. Write each order complete within itself, so as to have no reference to any former transaction.

When a customer fails to give shipping instructions, we forward good by express, rather than delay or write for advice.

We cannot hold ourselves responsible for goods that may be lost or damaged in the mails, either from or to us.

Should any of our customers have cause for complaint, we shall deem it a favor if they will state particulars, that we may thoroughly investigate, and do justice to all concerned.

Goods may be exchanged at purchaser's risk and expense, when returned promptly and in good condition.

We insure against loss in the mail, to any point within the United States, at the small cost of 5 cents, for any amount less than $5.00, and 5 cents extra for each additional $5.00, as we do not guarantee safe delivery if sent by mail.

Express is frequently cheaper, and always safe.

In returning goods by mail to be exchanged, please be careful to **erase** all marks, otherwise full letter postage is charged   The name and address of sender are allowed on the package, but writing matter inside also subjects it to letter postage.

Packages must be **tied,** so as to admit of being easily opened by the Postmaster for inspection.

To secure a fit it is desirable in ordering to give us the width as well as the length. Shoes are generally marked on the lining in figures to denote the length, and in letters to denote the width.   In case of doubt, any shoe which has been worn, however worthless it may be, will materially assist us in deciding the size.

Children generally require with each new pair of shoes at least half a size longer than they have been wearing.

Please do not cut out the illustrations.   You will be understood by quoting the number of the article wanted.

We will be pleased to reply to inquiries respecting goods, prices, styles, etc., endeavoring to give full information.

All communications should be addressed to

A. ALEXANDER,

Sixth Avenue and 19th Street,

NEW YORK CITY.

CUSTOM DEPARTMENT.

꙳ ꙳

We guarantee a fit to our own measurement only.

Measures left with us can be used for subsequent orders, but due consideration should be given to the fact, if the weight of the person has materially changed.

HOW TO SELECT.

Care.—There is no part of a lady's or gentleman's outfit that requires more intelligent care than the footwear, and, as a rule, none receives less.

Choice.—Principles of selection (always important), and in other articles of dress generally applied, are, in the choice of shoes, frequently overlooked. Sometimes both comfort and service are sacrificed to appearances. Having decided what is wanted, choose according to the use for which the goods are designed.

Economy.—Sometimes fine goods are chosen, when heavier goods would give much better satisfaction, besides being less expensive. Light fine shoes (like light fabrics) cannot stand so much hard usage as the heavier grades.

HOW TO FIT.

Guidance.—Our effort is to furnish a good fit. This is to our interest as well as that of the wearer. Insist on being properly fitted. The letter (indicating the width), and the number (indicating the length) of former shoes which have fitted satisfactorily, together with a description of their style and form, is a safe guide. Or, better still, forward by mail any shoe which has been worn long enough to have acquired the shape of the foot. We can then discover the peculiar characteristics (if any) required to secure a fit.

Length.—The form of the foot varies, perhaps more than that of any other portion of the body; so many feet have been distorted by poorly fitting shoes, especially by wearing shoes too short while the young feet are still growing. Do not attempt to wear short shoes. Test the length by standing up (full weight); if then there be any pressure on the toe, however slight, the shoe is too short.

Closeness.—Let them fit closely over the instep, so as to prevent their riding forward and throwing the pressure on the toes.

Exchanges.—Exchange rather than begin to wear poorly fitting shoes; but do not ask us to exchange shoes that have been soiled from use, or in which alterations have been made. Exchangeable goods may be exchanged at purchaser's risk and expense when returned promptly and in good condition.

Service.—Shoes that fit well not only look better, but also wear better than poorly fitting ones. Undue or wrongly distributed strain distorts the shape of both shoes and feet.

Style.—Much is said about style of lasts in securing accuracy of fit, symmetry, etc., etc. Each style may have special merits for its own type of foot, but no one style has universal claims. Do not follow the changes of style too closely. In many cases do not follow them at all.

Self=Measurement.—It is absurd.

The Illustrations are intended to show the **general** style. It would take many more than the number we now publish to show all the little modifications.

HOW TO WEAR.

Begin Right.—In putting on new shoes, be careful tor the first few times to have them perfectly straight on the foot. In buttoned shoes be quite as careful the first few times buttoning. More attention should be given to this than the first putting on of kid gloves. A button-hook used as a lever may break any button-hole, however strong. A little attention just here would save much trouble and obviate many misunderstanding.

Begin Right.—Do not handle or put on **Patent Leather** Shoes, especially new ones, while they are cold. Any grade of patent leather will crack readily when chilled.

Friction.—Wear of the uppers must always be expected where they are exposed to friction. "Constant dropping wears away a stone." When the rub comes from a harsh substance, and is applied to a single spot, as happens in many boyish and childish amusements, and in the case of some of the facings used in ladies' skirts, the effect, of course, is damaging. Such things, if they cannot be prevented, should never be complained of. They are expected and accepted in other articles of merchandise.

Complaints.—Kindly allow us to investigate fully each case. Let us see both shoes. It is our interest, even more than that of the wearer, to discover any fault or flaw in stock or workmanship.

Accidents.—Never dry shoes hurriedly, as you may readily burn them. By placing shoes in contact with heated rails or steam pipes, or near a fire, while leather is soaked with water, the drying process is rapid and they become scorched or burnt in a short time. So-called accidents frequently happen in this way.

Perspiration.—Do not wear rubbers over good shoes. Their tendency is to cause perspiration, and perspiration is little less destructive than acid to most kinds of leather.

Repairs.—Do not let shoes run down at the heel or side. To prevent this have them promptly repaired. This is true economy. Besides, by being kept in good condition, they retain their normal shape. Do not allow a crust of blacking or dressing to form on shoes.

DEFINITION OF SIZES.

❧❧

Ladies' sizes are ordinarily from Nos. 2 to 8 ; but in fine dress wear we also have them as small as Nos. 1 and 1½.

Misses' sizes are from .... 11 to 2
Children's sizes are from.... .... 6 to 10½
Infants' sizes are from........................ ....1 to 7
Men's sizes are from............. 5 to 11, in some styles, 12
Boys' sizes are from............................ ....1 to 5½
Youths' sizes are from......................... 11 to 13½

DEFINITION OF WIDTHS.

❧❧

AA .... Extra Narrow
A...Narrow
B .... Medium Narrow
C ....Medium
D ....Wide
E .... Extra Wide
EE .... Widest Made
N ....Narrow
M..... ....Medium
F ....Full
FF.. ....Extra Full

GENTS' SHOES.

Sizes 5 to 12.   Widths AA to E.

CADET TOE.

C. 211. Best French Patent Leather, Foxed, Kid Top, Laced Boot, Tip, Medium Weight Sole, "Cadet" Last.................$7 00

C. 47. French Patent Leather, Foxed, Kid Top, Lace Boot, Tip, Medium Weight Sole, "Cadet" Last.......$5 00

C. 0410. Patent Leather, Foxed, Kid Top, Balmoral, Tip, Medium Sole, "Cadet" Last. $4 00

C. 81. English Enamel Leather, Foxed, Kid Top, Balmorals, Calf Lined, Tip, Double Sole, "Cadet" Last........$6 00

C. 11. English Enamel Leather, Foxed, Kid Top. Balmorals, Tip, Double Sole, "Cadet" Last.... ....$5 00

C. 225. Fine French Calf, Foxed, Kangaroo Top, Balmoral, Tip, either Single or Double Sole, "Cadet" Last...$6 00

C. 503. Fine Calf, Foxed, Kangaroo Top, Balmoral, Tip, Heavy Single Sole, "Cadet" Last.................$5 00

C. 354. Best Hamburg Cordovan, Foxed, Kid Top. Balmorals, Tip, Calf Lined, Double Sole, "Cadet" Last...$6 00

C. 55. Fine Calf, Foxed, Kangaroo Top, Balmoral, Tip, Heavy Single Sole, "Cadet" Last.................$4 00

C. 361. Fine Calf, Foxed, Kangaroo Top, Balmoral, Tip. Heavy Single Sole, "Cadet" Last.................$3 00

7

GENTS' SHOES.

Sizes 5 to 11.   Widths AA to E.

CADET TOE.

C. 246. Tan Russia Leather Balmorals, Tip, Stout Single Sole.
"Cadet" Last $6 00

C. 706. Brown "Cordovan" Balmorals, Tip, Double Sole, "Cadet"
Last..................... ....$7 00

C. 122. Russet Oil Grain Balmorals, Tip, Double Sole, "Cadet"
Last... ....$5 00

C. 418. Russet Oil Grain Balmorals, Tip, Double Sole, "Cadet"
Last....... ....$4 00

C. 232. Tan Russia Leather Bluchers, Tip, Stout Single Sole, "Cadet"
Last................................... ....$6 00

C. 182. Tan Russia Leather Balmorals, Tip, Heavy Single Sole,
"Cadet" Last................................ ....$4 00

C. 608. Tan Russia Leather Balmorals, Tip, Heavy Single Sole,
"Cadet" Last...................................... ....$6 00

C. 600. Tan Russia Leather Balmorals, Tip, Stout Single Sole,
"Cadet" Last...$6 00

8

GENTS' SHOES.

Sizes 5 to 11.   Widths AA to E.

"CADET."

"CORNELL."

C. 0212. Best French Patent Leather, Foxed, Kid Top, Buttoned Boots, Tip, Medium Weight Sole, "Cadet" Last.................$7 00

C. 5. French Patent Leather, Foxed, Kid Top, Buttoned Boot, Tip, Medium Weight Sole, "Cadet" Last...$6 00

C. 9. French Patent Leather, Foxed, Kid Top, Buttoned Boot, Tip, Medium Weight Sole, "Cadet" Last..$5 00

C. 0237. Fine French Calf, Foxed, Kid Top, Buttoned Boot, Tip, "Cadet" Last, Heavy Single Sole......... $6 00

C. 0409. French Patent Leather, Foxed, Kid Top, Buttoned Boot, Tip, "Cadet" Last, Single Sole..........$4 00

C. 244. Best French Patent Leather, Foxed, Kid Top, Buttoned Boot, Tip, Medium Weight Sole, "Cornell" Last.................$7 00

C. 409. French Patent Leather, Foxed, Kid Top, Buttoned Boot, Tip, Single Sole, "Cornell" Last......$4 00

C. 047. French Patent Leather, Foxed, Kid Top, Buttoned Boot, Tip, Medium Weight Sole, "Cornell" Last..$6 00

C. 26. French Patent Leather, Foxed, Kid Top, Buttoned Boot, Tip, Single Sole, "Cornell" Last......$5 00

C. 237. Fine French Calf, Foxed, Kid Top, Buttoned Boot, Tip, "Cornell" Last, Heavy Single Sole............$6 00

9

GENTS' SHOES.

Sizes 5 to 11.   Widths AA to E.

"CORNELL."

C. 233. Best French Patent Leather, Foxed, Kid Top, Lace Boot, Tip, Medium Weight Sole, "Cornell" Last......................$7 00

C. 048. Best French Patent Leather, Foxed, Kid Top, Lace Boot, Tip, Medium Weight Sole, "Cornell" Last......................$6 00

C. 49. French Patent Leather, Foxed, Kid Top, Lace Boot, Tip, Medium Weight Sole, "Cornell" Last......................$5 00

C. 410. Patent Leather, Foxed, Kid Top, Lace Boot, Tip, Medium Weight Sole, "Cornell" Last..........................$4 00

C. 366. Vici Kid Balmoral, Tip, Single Sole, "Cornell" Last....$3 00

C. 602. English Enamel Leather, Foxed, Kid Top, Balmorals, Tip, Double Sole, "Cornell" Last $6 00

C. 371. Calf, Foxed, Dongola Top, Balmoral, Tip, Stout Single Sole, "Cornell" Toe......................................$2 00

C. 363. Fine Calf, Foxed, Kangaroo Top, Balmoral, Tip, Heavy Single Sole, "Cornell" Last.......................................$3 00

C. 56. Fine Calf, Foxed, Kangaroo Top, Balmoral, Tip, Heavy Single Sole, "Cornell" Last..$4 00

C. 216. Fine Calf, Foxed, Kangaroo Top, Balmoral, Tip, Heavy Single Sole, "Cornell" Last.......................................$5 00

C. 217. Fine French Calf, Foxed, Kangaroo Top, Balmoral, Tip, Heavy Single Sole, "Cornell" Last...............................$6 00

C. 236. Fine French Calf, Foxed, Kangaroo Top, Balmorals, Tip, Calf Lined, Double Sole, "Cornell" Last..................$6 00

10

GENTS' SHOES.

Sizes 5 to 11.   Widths AA to E.

"CORNELL."

C. 412. Tan Russia Balmorals, Tip, Heavy Single Soles, "Cornell" Last......$3 00

C. 330. Heavy Tan Russia Balmorals, Tip, Double Sole, "Cornell" Last......$3 00

C. 234, Tan Russia Leather Balmoral, Tip, Heavy Single Sole, "Cornell" Last .$6 00

C. 0123. Tan Russia Leather Balmorals, Tip, Heavy Single Sole, "Cornell" Last..$5 00

C. 302 to C. 307.

C. 304. Calf Balmorals, Cork Sole, London Toe, Tip............$5 00
C. 305. Calf Balmorals, Cork Sole, Medium Toe, Plain.......... 5 00
C. 307. Calf Balmorals, Cork Sole, London Toe, Tip............ 7 00
C. 302. Calf Balmorals, Cork Sole, Medium Toe, Plain.......... 9 00
C. 303. Calf Balmorals, Cork Sole, London Toe, Tip..... .... 9 00

GENTS' SHOES.

Sizes 5 to 11. Widths AA to E.

"LONDON."

C. 411. Tan Russia Leather Balmorals, Tip, Heavy Single Soles, "London" Toe $3 00

C. 181. Tan Russia Leather Balmorals, Tip, Heavy Single Sole, wide London Toe... ....$4 00

C. 184. Tan Vici Kid Balmorals, London Toe, Tip, Single Sole..$5 00

C. 100. Russet Sealskin Balmorals, Tip, Heavy Single Soles, London Toe $6 00

C. 105. Tan Russia Leather Balmorals, Tip, Heavy Single Soles, wide London Toe $6 00

C. 607. Tan Russia Leather Balmorals, Tip, Heavy Single Sole, wide London Toe $6 00

GENTS' SHOES.

Sizes 5 to 12.   Widths AA to E.

C. 217.

"LONDON" TOE.

C. 370. Calf, Foxed, Dongola Top Balmoral, Tip, Stout Single Sole, London Toe.....$2 00

C. 362. Calf, Foxed, Dongola Top Balmorals, Tip, Stout Single Sole, London Toe.....$3 00

C. 77. Fine Calf, Foxed, Kangaroo Top, Bal., Tip, Heavy Single Sole, London Toe, $4 00

C. 51. Best Hamburg Cordovan Balmoral, Tip, Heavy Single Sole, London Toe..$6 00

C. 58. Fine French Calf, Foxed, Kangaroo Top, Bal., Tip, Hand Sewed, Med. Weight Sole. "London" Toe.$7 50

C 214. Fine French Calf, Foxed, Kangaroo Top, Balmoral, no Tip, Hand Sewed, Medium Sole, "London" Toe..$7 50

C. 403. English Enamel Leather, Foxed, Kid Top, Balmorals, Tip, Double Sole, London Toe $4 00

C. 241. Best Hamburg Cordovan, Foxed, Kid Top, Balmorals, Tip, Double Sole, wide London Toe. Calf Lined ..$7 00

C. 704. Fine French Calf, Foxed, Kangaroo Top, Balmoral, Tip, Double Sole, London Toe, Calf Lined......$7 00

C. 74. Fine French Calf, Foxed, Kangaroo Top, Balmoral, Tip, Single Sole, Bevel Edge, "London" Toe.......$6 00

C. 0224. English Enamel Leather, Foxed, Kid Top, Balmorals, Single or Double Sole, Narrow London Toe......$6 00

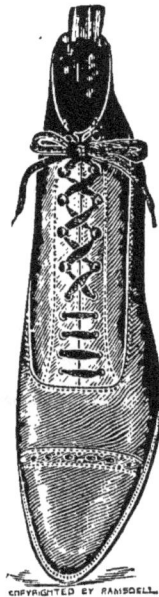

"OPERA" TOE.

C. 63. Fine Calf, Foxed, Kangaroo Top, Balmoral, Tip, Light Weight Sole, "Opera" Toe......... ....$5 00

C. 53. Fine Cordovan Foxed, Kangaroo Top, Bal., No Tip Single Sole, "Opera" Toe, $5 00

C. 223. Best Hamburg Cordovan, Foxed, Kangaroo Top, Balmoral, Tip, Single Sole, Opera Toe...........$6 00

C. 062. Fine Vici Kid Balmoral, Tip, Single Sole, "Opera" Toe $5 00

C. 218. French Pat. Leather, Foxed, Kid Top, Balmoral, No Tip, Single Sole, Bevel Edge, Narrow London Toe..$6 00

C. 7. French Patent Leather, Foxed, Cloth Top, Balmoral, No Tip, Single Sole, Narrow London Toe..........$6 00

C. 19. French Patent Leather, Foxed, Kid Top, Balmoral, Tip, Single Sole, Opera Toe, $5 00

C. 0210. French Patent Leather, Foxed, Kid Top, Balmoral, Tip, Single Sole, Narrow London Toe..........$6 00

C. 229. French Patent Foxed, Kid Top, Balmoral, Tip, Single Sole, London Toe.... $6 00

GENTS' SHOES.

Sizes 5 to 11.   Widths AA to E.

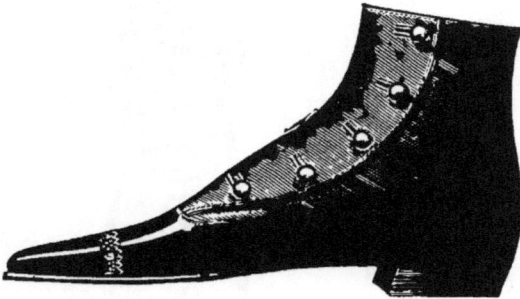

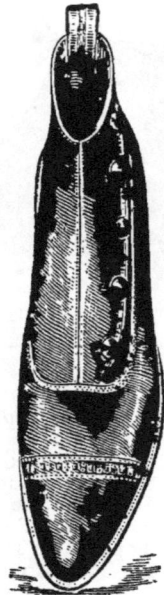

C. 0229.

C. 0229. French Patent Leather, Foxed, Kid Top, Buttoned Boot, Tip, Single Sole, "Opera" and "London" Last.....$6 00

C. 3.   French Patent Leather, Foxed, Kid Top, Buttoned Boot, Tip, Single Sole, "Opera" and "London" Last, Narrow London Toe $5 00

C. 140.   French Patent Leather, Foxed, Cloth Top, Buttoned Boot, Tip, Single Sole, Narrow London Toe. $6 00

C. 68.   Fine Calf, Foxed, Kid Top, Buttoned Boot, Tip, London Toe, Single Sole.. ....$5 00

C. 66.   Fine French Calf, Foxed, Kid Top, Buttoned Boot, No Tip London Toe, Hand Sewed, Single Sole...... ....$7 00

C. 65.   Fine French Calf, Foxed, Kid or Cloth Top, Buttoned Boot, Tip, London Toe, Hand Sewed, Single Sole $7 00

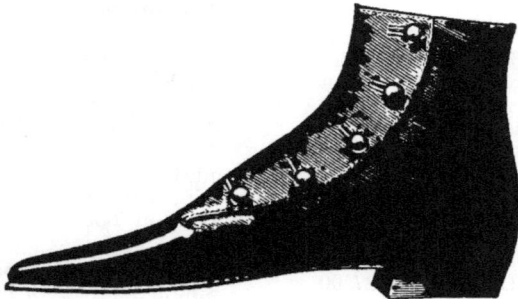

C. 188.   French Patent Leather, Foxed, Cloth Top, Buttoned Boot, No Tip, Single Sole, Narrow London Toe................ .....$6 00

C. 4.   French Patent Leather, Foxed, Kid Top, Buttoned Boot, No Tip, Single Sole, French Toe............. ....$6 00

14

GENTS' SHOES.

Sizes 5 to 11.   Widths AA to E.

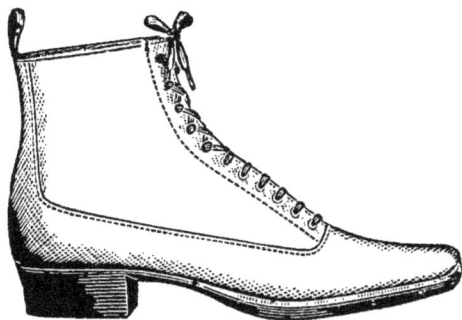

"FRENCH" TOE.

C. 52.   Fine Calf, Foxed, Kangaroo Top, Bal., No Tip, Single Sole, French Toe..$5 00

C. 78.   Fine Calf, Foxed, Kangaroo Top, Bal., No Tip, Single Sole, French Toe..$4 00

C. 345.   Fine Calf, Foxed, Kangaroo Top, Bal., No Tip, Single Sole, French Toe..$3 00

C. 705.   Fine French Calf, Foxed, Kangaroo Top, Balmoral, No Tip, Calf Lined, Double Sole, French Toe $7 00

C. 213.   Fine French Calf, Foxed, Kangaroo Top, Balmoral, No Tip, Hand Sewed, Medium Sole, French Toe.$7 50

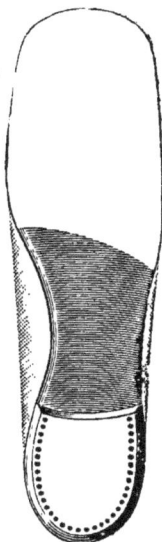

"FRENCH" TOE.

C. 6.   French Patent Leather, Foxed, Kid Top, Balmoral, Single Sole, French Toe.$6 00

C. 91.   Fine Vici Kid Balmoral, No Tip, Single Sole, Medium French Toe..........$5 00

C. 334.   Vici Kid Balmoral, No Tip, Single Sole, French Toe.................$3 00

C. 101.   Russet Sealskin Balmorals, No Tip, Heavy Single Sole, Full French Toe.................$6 00

C. 369.   Fine Calf, Foxed, Kangaroo Top Balmorals, Tip, Calf Lined, Cork Sole, French Toe, Sizes 6 to 12................ ......$3 50

C. 368.   Fine Calf, Foxed, Kangaroo Top, Balmorals, Tip, Calf Lined, Double Sole, French Toe, Sizes 6 to 12....................$3 50

These two lines are especially adapted for Policemen, Firemen, and Letter Carriers.

GENTS' SHOES.

Sizes 5 to 11.   Widths AA to E.

C. 67.   Fine Calf Foxed, Kid Top, Congress Gaiter, No Tip, Medium French Toe.. .....$4 00

C. 71.   Fine French Calf Foxed, Kid Top, Congress Gaiters, Hand Sewed, No Tip, wide London Toe..................... .... ..$7 00

C. 72.   Fine Calf Foxed, Kid Top, Congress Gaiter, No Tip, French Toe.. ....$5 00

C. 61.   Fine Vici Kid Congress, No Tip, Single Sole, Medium Toe, $3 00 and $5 00

C. 1.   Patent Leather, Foxed, Satin de laine Top, Congress Gaiters, London Toe, No Tip, " Full Dress Shoes".................$5 00

C. 10.   Patent Leather, Foxed, Kid Top, Congress Gaiters, London Toe, No Tip $5 00

C. 300.   Calf Congress, Cork Sole, Toe Plain....................$7 00

C. 301.   Calf Congress, Cork Sole, London Toe, Tip............. 7 00

C. 73.   Fine Calf Foxed, Kid Top, Con. Gaiters, Tip, London Toe..$5 00

C. 341.   Fine Calf, Foxed, Kid Top, Congress Gaiter, Tip, London Toe, Heavy Single Sole..$3 00

C. 342.   Fine Calf, Foxed, Kid Top, Congress Gaiter, Tip, Medium French Toe, Heavy Single Sole.......... ... .$3 00

C. 2.   Patent Leather, Foxed, Kid Top, Congress Gaiters, Tip, London Toe..$6 00

16

GENTS' SHOES.

Sizes 5 to 11.   Widths AA to E.

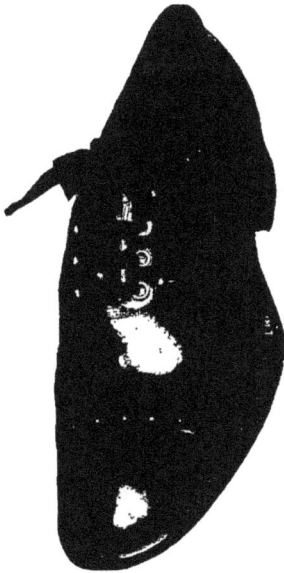

C. 415. French Patent Leather Oxford Ties, Tip, Single Sole, "Cadet" Last...$4 00

C. 0219. Best French Patent Leather Oxford Ties, Tip, Heavy Single Sole, "Cadet" Last................$6 00

C. 134. Fine French Calf Oxford Ties, Tip, "Cadet" Last, Single Sole $5 00

C. 414. Tan Russia Oxford Ties, Tip, Heavy Single Sole, "Cornell" and "Cadet" Last........ ....$3 00

C. 604. Heavy Russia Leather Oxford Ties, Tip, Full Double Sole, "Cadet" Last................ $6 00

C. 0235. Tan Russia Oxford Ties, Tip, Heavy Single Sole, "Cadet" Last.... ...$5 00

C. 14. French Patent Leather Oxford Ties, Tip, Heavy Single Sole, "Cornell" Last.............. ...$5 00

C. 243. Fine Calf Oxford Ties, Tip, "Cornell" Last, Single Sole. $3 00

C. 132. Fine Calf Oxford Ties, Tip, "Cornell" Last, Single Sole $4 00

C. 219. Best French Patent Leather Oxford Ties, Tip, Heavy Single Soles, "Cornell" Last...........$6 00

C. 29. Tan Russia Oxford Ties, Tip, Heavy Single Sole, "Cornell" Last......$4 00

C. 245. Tan Russia Oxford Ties, Tip, Heavy Single Soles, "Cornell" Last.......$5 00

17

GENTS' SHOES.

Sizes 5 to 11.   Widths AA to E.

C. 118.   Fine French Calf Oxford Ties, No Tip, London Toe, Pump
Sole....... ....$4 50

C. 176.   Dongola Oxford Ties, No Tip, London Toe, Single Sole, abso-
lutely noiseless, an excellent Shoe for waiters...........$2 00

C. 367.   Vici Kid Oxford Ties, No Tip, French Toe, Single Sole...$3 00

C. 12.   French Patent Leather Oxford Ties, No Tip Single Sole,
French Toe............... ....$5 00

C. 116.   Fine French Calf Oxford Ties, No Tip, French Toe, Hand
Sewed, Single Sole............................. ....$6 00

C. 107.   Russet Sealskin Oxford Ties, No Tip, Single Sole, French
Toe................................. ....$5 00

C. 117.   Fine French Calf Oxford Ties, Tip, London Toe, Hand Sewed,
Single Sole. $6 00

C. 222.   French Patent Leather Oxford Ties, Tip, Single Sole, " Opera "
Toe............... ....$5 00

C. 13.   French Patent Leather Oxford Ties, Tip, Single Sole, London
Toe..... ....$5 00

C. 91.   Vici Kid Oxford Ties, Tip, London Toe, Single Sole...... 4 00

C. 106. Russet Sealskin Oxford Ties, Tip, Single Sole, London Toe, 5 00

BOYS' SHOES.

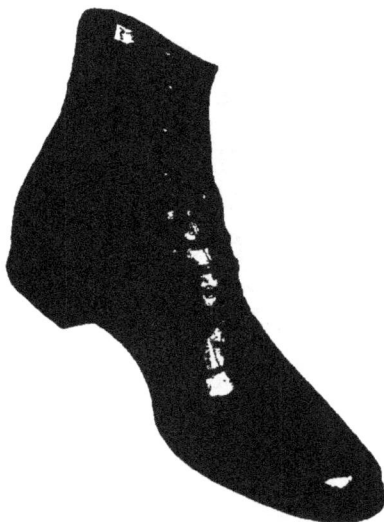

C. 34. Fine Calf Foxed, Kangaroo Top, Balmoral, Tip, London Toe,
    Stout Single Sole, Sizes, 11 to 13½..... .. ..$2 50

    Sizes, 1, 1½, and 2........................ .... 2 75

    Sizes, 2½ to 4½....... .. .. 3 50

    Sizes, 5 and 5½. 4 00

C. 37. Fine Calf Foxed, Kangaroo Top, Balmoral, Tip, London Toe,
    Heavy Single Sole, Sizes, 2½ to 5½.....................$3 00

C. 39. Fine Calf Foxed, Kangaroo Top, Balmorals, Tip, Double Sole,
    "London" Toe, Sizes, 1 to 4½............................$2 75

    Sizes, 5 and 5½ 3 00

C. 44. Calf Foxed, Kangaroo Top, Balmorals, Tip, "Steel Shod,"
    Double Sole, Sizes, 1 to 5½.......................... ....$2 00

C. 119. Calf Foxed, Kangaroo Top, Balmorals, Tip, "Steel Shod,"
    Double Sole, Sizes,11 to 13½ $1 25

    Sizes, 1 to 5½............. .... 1 50

C. 24. Patent Leather Foxed, Kid Top, Balmorals, Tip, Single Sole,
    London Toe, Sizes, 11½ to 2................. ....$3 00

    Sizes, 2½ to 4½.. 3 50

    Sizes, 5 and 5½........... .... 4 00

C. 0109. Tan Russia Balmorals, Tip, Stout Single Sole, London Toe,

    Sizes, 11 to 13½.............. ....$2 75

    Sizes, 1, 1½ and 2........................... .... 3 00

    Sizes, 2½ to 4½............... ... . 3 50

    Sizes, 5 and 5½ 4 00

C. 45. Tan Russia Balmorals, Stout Single Sole, London Toe, Tip,
    Sizes, 11 to 2........ ....$1 75

    Sizes, 2½ to 5½... .... 2 00

BOYS' SHOES.

Sizes 11 to 2, 2½ to 5½. Widths AA to E.

C. 143.   Patent Leather Foxed, Kid Top, Balmorals, Tip, Single Sole, "Cadet" Last, Sizes, 1, 1½ and 2. ....$3 00
Sizes, 2½ to 4½....$3 50   Sizes, 5 and 5½........... 4 00

C. 379.   English Enamel Leather Foxed, Kid Top, Balmorals, Tip, Heavy Single Sole, "Cornell" Last, Sizes, 1, 1½ and 2.....$3 00
Sizes, 2½ to 5½... ....$3 50

C. 336.   Vici Kid Balmorals, Tip, Heavy Single Sole, "Cornell" Last, Sizes, 1 to 2................ ... .$3 00
Sizes, 2½ to 5½... 3 50

C. 0242.   Fine Calf Foxed, Kangaroo Top, Balmorals, Tip, Full Double Sole, "Cornell" Last, Sizes, 2½ to 5½....................$4 00

C. 87.   Fine Calf Foxed, Kangaroo Top, Balmorals, Tip, Full Double Sole, "Cadet" Last, Sizes, 2½ to 4½... ....$3 50
Sizes, 5 and 5½....................................... 4 00

C. 79.   Fine Calf Foxed, Kangaroo Top Balmorals, Tip, Double Soles, "Cornell" Last, Sizes, 2½ to 5½......... ....$3 00

C. 126.   Tan Russia, Balmorals, Tip, Stout Single Sole, "Cornell" Last, Sizes, 2½ to 5½.................. ... .$3 00

C. 108.   Tan Russia, Balmorals, Tip, Stout Single Sole, "Cadet" Last, Sizes, 1, 1½ and 2.....................................$3 00
Sizes, 2½ to 4½ 3 50
Sizes, 5 and 5½............................... .... 4 00

BOYS' SHOES.

C. 139.   Tan Oil Grain Balmorals, Tip, Full Double Sole, "Cadet"
Last, Sizes, 2½ to 4½ $4 00
Sizes, 5 and 5½....................................... 4 50

C. 378.   Tan Russia, Balmorals, Tip, Full Double Sole, "Cornell" Toe,
Sizes, 2½ to 5½...$3 00

C. 046.   Tan Russia Balmorals, Tip, Heavy Single Sole, "Cornell"
Last, Sizes, 11 to 2.......................................$2 25
Sizes, 2½ to 5½........................... .... 2 50

C. 46.   Fine Calf Foxed, Kangaroo Top, Balmorals, Tip, Opera Toe,
Heavy Single Sole, 11 and 2 .. ..$2 25
2½ to 5½ 2 50

C. 203.   Calf Foxed, Kangaroo Top, Balmorals, Tip, Opera Toe, Heavy
Single Sole, 1 to 5½......... ... .$2 00

Sizes 2½ to 5½.

C. 20.   Patent Leather Oxford Ties, Tip, Single Sole, "Cornell"
Toe, Sizes, 2½ to 5½.....................................$4 00

C. 114.   Fine Calf Oxford Ties, Tip, Single Sole, "Cornell" Last,
Sizes, 2½ to 4½...$3 00
Sizes, 5 and 5½ 3 50

C. 113.   Tan Russia Oxford Ties, Tip, Single Sole, "Cornell" Last,
Sizes, 1 to 4½ $3 00
Sizes, 5 and 5½ 3 50

C. 248.   Tan Russia, Oxford Ties, Tip, Single Sole, "Cadet" Last,
Sizes, 1 to 4½...$3 00
Sizes, 5 and 5½ 3 50

C. 115.   Tan Russia Oxford Ties, Tip, Single Sole, London Toe,
Sizes, 1 to 2...... ....$2 25
Sizes, 2½ to 5½.. 2 50

SPORTING GOODS.

C. 86 to C. 202.

C. 86. Gent's Black Box Calf Bicycle Balmoral, Elk Sole, London Toe..$3 00

C. 83. Gent's Tan Box Calf Bicycle Balmoral, Elk Sole, London Toe..$3 00

C. 84. Gent's Tan Russia Leather Bicycle Balmoral, Corrugated Leather Sole, London Toe,..............................$2 50

C. 85. Gent's Black Calf Bicycle Balmoral Corrugated Leather Sole, London Toe.......................................$2 50

C. 88. Gent's White Canvas Bicycle Balmoral, Corrugated Leather Sole, London Toe... ....$2 50

C. 201. Gent's Black Calf Bicycle Balmoral, Corrugated Leather Sole, London Toe.................$2 00

C. 202. Gent's Russet Calf Bicycle Balmoral, Corrugated Leather Sole, London Toe $2 00

C. 94. Gent's Russet Calf Bicycle Balmoral, Corrugated Leather Sole, Tip, London Toe.................$3 00

C. 83. Boy's Tan Box Calf Bicycle Balmoral, Elk Sole, London Toe, Sizes 2½ to 5½................. $2 50

C. 94.

C. 84. Boy's Tan Russia Bicycle Balmoral, Corrugated Leather Sole, London Toe, Sizes 13 to 4½..........................$2 25

C. 85. Boy's Black Calf Bicycle Balmoral, Corrugated Leather Sole, London Toe, Sizes 1 to 4½... ....$2 25

22

SPORTING GOODS.

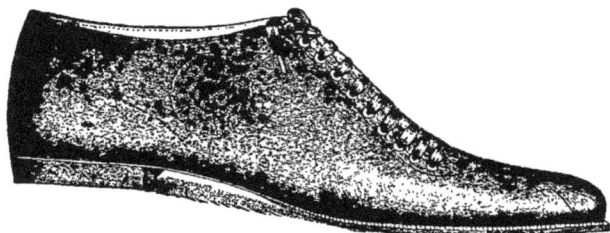

C. 82.

C. 82.   Gent's Tan Russia Bicycle Oxfords, Elk Sole, London Toe, $2 50

C. 95.   Gent's Black Calf Bicycle Oxfords, Elk Sole, London Toe, 2 50

C. 204. Gent's Russia Calf Bicycle Oxfords, Corrugated Leather Sole,
London Toe $2 00

C. 205. Gent's Black Calf Bicycle Oxfords, Corrugated Leather Sole,
London Toe.... ....$2 00

C. 130.   Gent's Tan Russia Ten-
nis Balmoral, Heavy Smooth
Rubber Sole, Tip, "Cadet"
Last.... ....$4 00

C. 129.   Gent's Tan Russia Ten-
nis Oxfords, Heavy Smooth
Rubber Sole, Tip, "Cadet"
Last..................$3 50

C. 128.   Gent's Russet
Goat Tennis Ox-
fords,   H e a v y
Smooth   Rubber
Sole, Tip, London
Toe $2 50

C. 130.

C. 125.   Boy's Tan Russia Leather Tennis Oxfords, Heavy Smooth
Rubber Sole, Tip. London Toe, Sizes 2½ to 5½..... .. ... $3 00
Sizes 11 to 2... 2 50

C. 127.   Boy's Tan Russia Leather Tennis Balmorals, Heavy Smooth
Rubber Sole, Tip, "Cadet" Last, Sizes 2½ to 4½...........$3 00
Sizes 5 and 5½................................... ....3 50

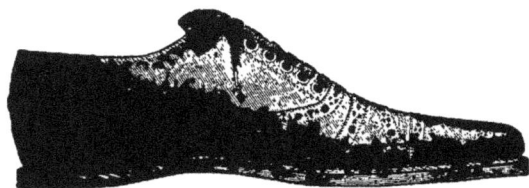

C. 128.

C. 133.   Gent's   White
Canvas   Tennis
Balmorals, Heavy
Smooth   Rubber
Sole, Tip, "Cor-
nell" Last...$4 00

C. 137.   Gent's   White
Canvas   Tennis
Oxfords,   Heavy
Smooth   Rubber

Soles, Tip, "Cornell" Last.............................$3 50

C. 135.   Gent's White Canvas Balmorals, Medium Stout Leather Sole,
Tip, "Cadet" Last... ....$4 00

23

SPORTING GOODS.

C. 136. Gent's White Canvas Oxford Ties, Medium Stout Leather Sole, Tip, "Cadet" Last................. ....$3 50

C. 335. Gent's Brown Canvas Balmoral, Medium Stout Leather Sole, Tip, "Cornell" Last...... ....$3 00

C. 137.

X. 920. Gent's White, Brown and Black Canvas Tennis Balmorals....$1 25

Boy's White, Brown and Black Canvas Tennis Balmorals....$1 25

Gent's White, Brown and Black Canvas Tennis Oxfords......$1 00

Boy's White, Brown and Black Canvas Tennis Oxfords,

Sizes 1 to 5½..$0 85

Sizes 9 to 13½...0 75

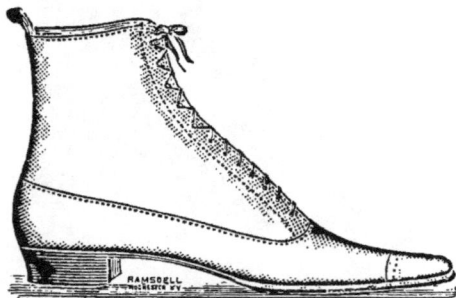

C. 135.

C. 684. Russet Grain Golf Balmorals, R u b b e r disks in Soles and heels, Double Sole.......$6 00

C. 501. Russet Grain Golf Balmorals, Double Sole hob nails in Soles and Heels............$5 00

Russet Grain Foot Ball Shoes.......$5 00

Russet Grain Base Ball Shoes, without Steel Clamps, $4 50

Russet Grain Base Ball Shoes, with Steel Clamps.. 5 00

Racing and Jumping Shoes, Steel Spikes....... .. 3 00

Bicycle Racing Shoes............................. 2 00

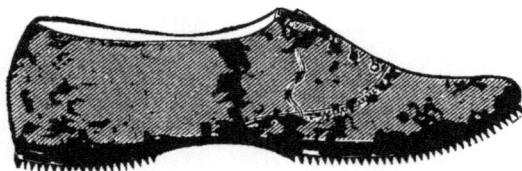

X. 920.

C. 604.

24

SPORTING GOODS.

C. 651.   Black and Tan Oil Grain Hunting Blucher, Double Sole, Extra
  High Cut.... ....$6 00

C. 652.   Tan and Black Oil Grain Hunting Bluchers, Extra High Cut,
  Rawhide Sole, Hob Nailed.... ....$7 00

**Please do not cut out the illustrations.   You will be understood
by quoting the number of article wanted.**

Pay no attention to published rules for self=measurement.

**This Catalogue represents such goods as are kept constantly in
stock.**

**In returning goods by mail, to be exchanged, please be careful
to erase all marks, otherwise full letter postage is charged.   The
name and address of sender are allowed on the package, but writing
matter inside also, subjects it to letter postage.**

Do not attempt to wear short shoes.

**Write each order complete within itself, so as to have no
reference to any former transaction.**

SPORTING GOODS.

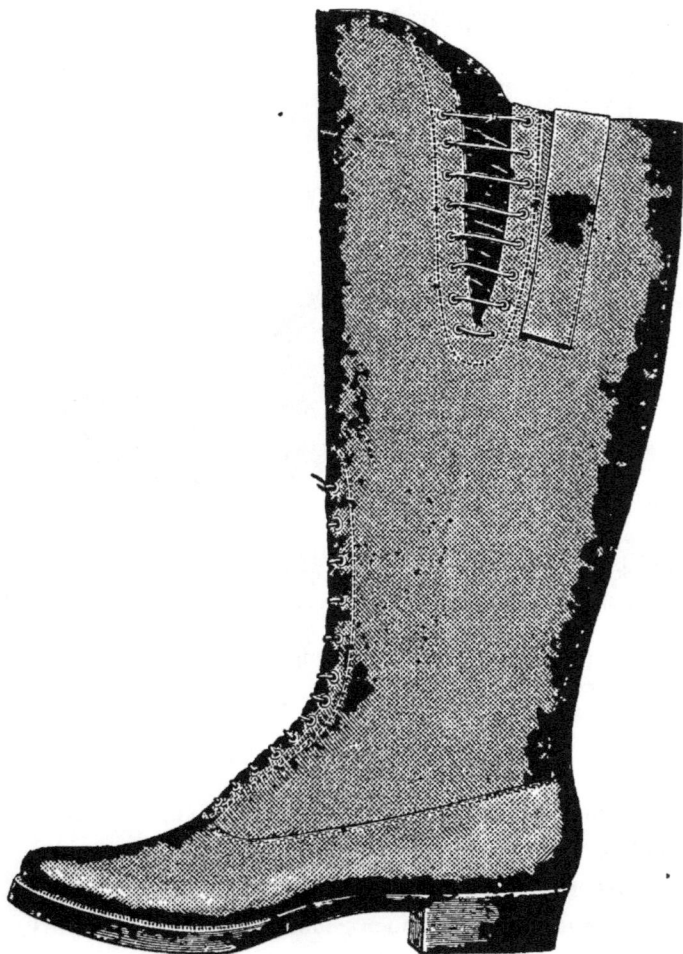

C. 801—802.

C. 801. Tan Oil Grain Hunting Boot, Rawhide Sole, Hob Nailed, $8 00

C. 802. Black and Tan Oil Grain Hunting Boot, Double Sole.... 8 00

This Catalogue represents such goods as are kept constantly in stock.

The quality of goods, with their accuracy of fit, is the true test of cheapness.

X. 918.

Gent's Fine American Calf Feet, Morocco Leg Boot.............$5 00
Gent's Fine French Calf Feet, Morocco Leg, Boot, Hand Sewed. $10 00

RUBBER GOODS. .

❧❧

Rubber Boots, Arctics and Over Shoes of all kinds, for Men and Boys. Only the *BEST* goods. In Over Shoes and Boots, as in other kinds of Footwear, the *BEST IS ALWAYS THE CHEAPEST*.

A. ALEXANDER,

Sixth Avenue and 19th Street.

GENTS' AND BOYS' SLIPPERS.

C. 156.

C. 156.   Russet and Black Seal Romeo Slipper...................$4 00

C. 157.   Russet Seal Romeo, Elastic Sides, London Toe.......... 4 00

C. 170.   Black Kid Romeo, Elastic Sides, London Toe........... 2 50

C. 171.   Russet Goat Romeo, Elastic Sides, London Toe......... 2 50

C. 177.   Russet and Black Romeo, Elastic Sides, London Toe.......$2 00

C. 168.   Boy's Russet Goat Romeo, Elastic Sides, London Toe, Sizes, 1 to 5½................$2 00

C. 153.   Gent's Russet Seal Opera Slipper, London Toe, Sizes, 5 to 12..$3 00

C. 154.   Gent's Brown Seal Opera Slipper, Sizes, 5 to 12..............$3 00

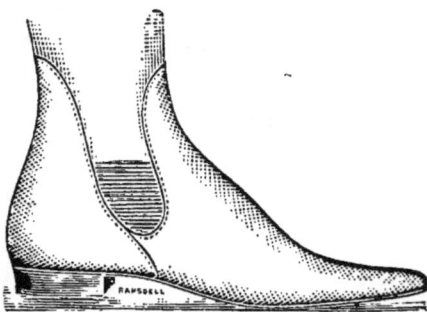

C. 170.

C. 155.   Gent's Black Seal Opera Slipper, Sizes, 5 to 12...........$3 00

C. 173.   Gent's Russet Goat Opera Slipper, Sizes, 5 to 12......... 2 00

C. 153.

C. 150.   Gent's Maroon Goat Opera Slipper, P. L. Trimmed, Sizes, 5 to 12... ....$2 00

C. 0173.   Gent's Russet Goat Opera Slipper, Sizes, 5 to 12........ 1 50

C. 0174.   Gent's Black Goat Opera Slipper, Sizes, 5 to 12......... 1 50

GENTS' AND BOYS' SLIPPERS.

C. 162.

C. 162.

C. 162. Boy's Russet and Maroon Opera Slipper. No Heel. Sizes, 6 to 8½..$1 00

Sizes, 9 to 13½.. 1 25

Boy's Russet and Maroon Opera Slipper, with Heels, Sizes, 1 to 4½..$1 75

Sizes, 5 and 5½......... .... 2 00

C. 16.

C. 16. Gent's Patent Leather Dress Pumps, Narrow London Toe.$2 50

C. 0351. Gent's Patent Leather Dress Pumps, " Cornell " Toe... 3 50

C. 195. Gent's Patent Leather Dress Pumps, " Cornell " Toe..... 3 00

C. 27.

C. 27. Boy's Patent Leather Oxford Pumps, Opera Toe, Sizes, 11 to 13½..... ....$2 00

Sizes, 1, 1½ and 2 .... 2 25

Sizes, 2½ to 5½...................... .... 2 50

C. 28.

C. 28. Boy's Patent Leather Dress Pumps, Opera Toe, Sizes, 11 to 13½ $2 00

Sizes, 1 to 4½.. 2 25

Sizes, 5 and 5½............... .... 2 50

C. 15. Gent's Patent Leather Oxford Pumps, Opera Toe......... 3 50

GENTS' RIDING BOOTS AND LEGGINGS.

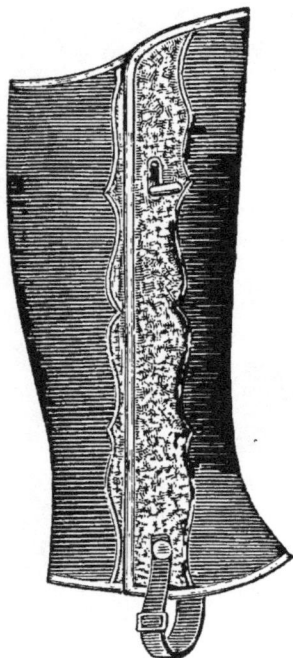

1201.

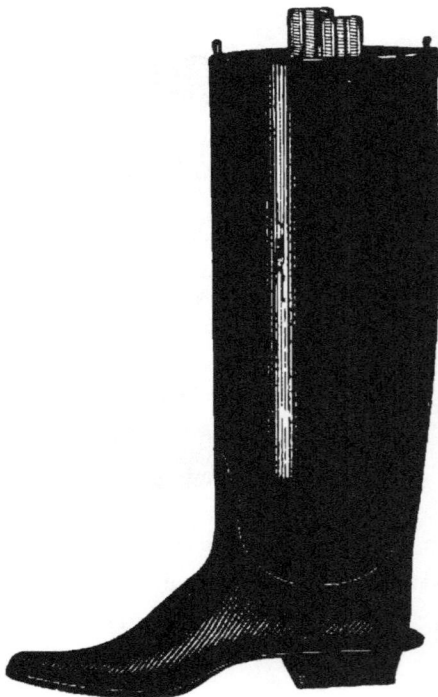

X. 919.

1201.   Gent's Tan Calf Riding Boots $12 00

Gent's Fine French Calf Riding Boots.............. .... 12 00

Gent's Fine French Calf Riding Boots........... .... 10 00

Regulation U. S. Officer's Calvary Boot..... .... 12 00

X. 919.   Black and Tan Grain Leather Riding Leggings..................$4 00

Brown Canvas Riding Leggings,
$2 00

Brown Canvas Shooting Leggings,
$1 50

Brown Canvas Bicycle Leggings,
$0 75

Black Canvas Bicycle Leggings,
$0 75

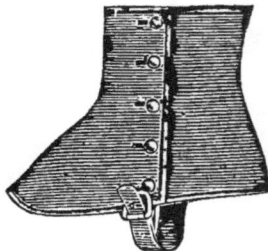

X. 917.

Gent's Imported Box Cloth Overgaiter................ .. ..$2 50

Gent's Tan Cloth Overgaiter............. .... 1 25

Gent's Black Cloth Overgaiter..............75c., $1 00, $1 25 and 1 50

Gent's Black Cloth Overgaiter, High Cut, Extra Quality.......... 2 00

FELT SLIPPERS.

Felt Uppers—Felt Soles.

Felt Uppers—Leather Soles.

Style 101. Nullifier.

Of fine black Toilet Felt. Perfection for a house or office shoe, and as easily put on as a slipper. It thoroughly protects the instep and ankle, and is just what is wanted by those who cannot wear slippers. Thick Felt Sole. Sizes, 5 to 12, Widths, B to EE. Price, $3 50

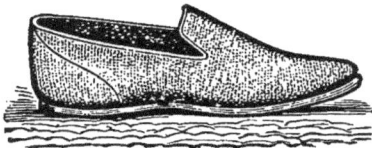

Style 90. Toilet, No Heel.

The standard Slipper, and is one of our most popular styles, made of Black Felt. Is shapely, comfortable and light. Widths, A to EE, Sizes, 5 to 12. Price, $1 50

No. 40.

Moulded from one piece of Toilet Felt without any sole.

Style 50. Rubber Boot Slipper.

They are worn inside of rubber boots to absorb moisture. All sizes, for Men, Boys and Ladies. Price .... 25 Cents.

Style 241. Romeo.

This slipper is made of one piece of Black Felt. The sole is belting leather—the most flexible and best wearing. The Romeo style is particularly designed for those who cannot wear low slippers without taking cold. Sizes, 5 to 12, Widths, B to E. Price..$1 50

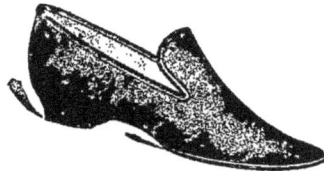

Style 141. Slipper.

This is made as just described only lower cut. Both Slippers have common sense toe. Sizes, 5 to 12, Widths, B to E. Price,$1 50

Style 188. "Comfort."

This is certainly the softest and lightest bedroom Slipper ever made. The upper is of a downy wool material without seams and without binding (patented). The sole is a light fine leather sole. Widths, B to E. Colors, Drab, Brown and Red. Men's sizes, 6 to 11. Price.............$1 25

31

SUNDRIES.

Shoe Trees in these days are almost a necessity to every lady and gentlemen wearing Fine Shoes. They keep the Shoes in shape, taking out wrinkles, and preventing the toes from turning up.

We keep them in different varieties, and make them to order to fit perfectly, either Shoes or Riding Boots.

We particularly recommend, on account of its compactness and convenience for handling, the

PUTTING TREE IN SHOE.

"PACK FLAT" SHOE TREE,

For Ladies' and Gentlemen's Shoes.

Keeps the Shoes in perfect shape, taking out all wrinkles and straightening out the sole.

Adjustable to any length of shoes.

Takes up no space.

Ventilated to allow shoes to dry inside, keeping it straight in the meantime.

TREE IN PLACE.

Takes the place of and is an improvement and far cheaper than the cumbersome old tree in three pieces.

Invaluable to travellers.

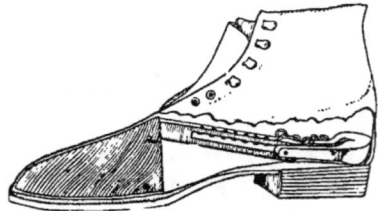

All who desire to have their shoes look nice should have a pair.

The Only Adjustable Shoe Tree on the Market.

ONCE ADJUSTED, ALWAYS ADJUSTED. LASTS A LIFETIME.

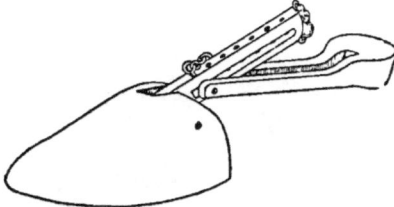

THE TREE. "

We also keep SHOE BLACKINGS and DRESSINGS of all kinds, for either Black or Brown Shoes.

SHOE BRUSHES, POLISHERS, PIPE CLAY FOR WHITE SHOES, and a general assortment of such "Sundries."

32

M. O. D.

.... *189*

A. ALEXANDER, Shoes,

Sixth Avenue and 19th Street, New York.

Please forward to M.............. ....

Use prefix Mr., Mrs., or Miss. Married ladies should invariably use husbands' initials.

Amt. Enclosed.

Address: Town.................. ... . *P. O. Order $*.

County .... *Express Order $*.............

State.. *Bank Draft* ....

How to Ship, Mail, or Express... *Cash $* . ...

Catalogue No. ....:.. .. ..

ze in logue	No. in Catalogue	DESCRIPTION.	Size.	Width.	Price.

www.ingramcontent.com/pod-product-compliance
Lightning Source LLC
Chambersburg PA
CBHW021452300326
41935CB00048B/644